Our Sun, Our Weather

Nancy White

Contents

Here comes the sun! As the sun rises, a new day begins.

The sun warms and heats
the land, water, and air.

1

2

The sun's warmth helps make our **weather.**

We get information about the weather from **meteorologists.** These scientists **predict** what the weather will be.

3 **4** **5**

| Monday | Tuesday | Wednesday | Thursday | Friday |

Look at this five-day weather **forecast.** Which photograph matches the forecast for each day?

On some days, the weather is warm and sunny.

Zion National Park, Utah

In some places, it can be very hot. The **temperature** rises very high!

The sun's warmth makes water **evaporate** into the air as **vapor.** The water vapor may cool to form tiny droplets. The droplets become clouds.

Hoh Rain Forest, Washington

When the droplets get big and heavy, they fall to the ground as rain.

In some places, it rains for many days in a row.

In very cold weather, water freezes and then falls as sleet or snow. Sometimes snow can be very deep!

Our weather is always changing. What prediction can you make about the weather in this photograph?

Even on cloudy days, the sun is still there, shining brightly above the clouds.

The sun helps to make wind, too. As the sun warms the land and water, it causes the air above to move.

What is happening on
this windy day?

At the end of each day, the sun sets. The temperature cools. On clear nights, we can see the moon and stars.

In the morning, another day will begin. Once more the sun will warm the land, water, and air.

Glossary

evaporate (ih-VAP-uh-rayt): dry up

forecast (FOR-kast): a statement of what will happen

meteorologist (mee-tee-uh-ROL-uh-jist): a scientist who studies, and may forecast, the weather

predict (prih-DIKT): tell what may happen in the future

temperature (TEM-pur-uh-chur): a measure of hotness or coldness

vapor (VAY-pur): a gas, such as the form of water that you cannot see in the air

weather (WEH-<u>th</u>ur): a day's hotness or coldness, wetness or dryness, calmness or windiness, and clearness or cloudiness